My Tornado Book

By Nicholas J. D. Sims

Copyright © 2023 by Nicholas J.D. Sims

All rights reserved. This book or any portion thereof may not be reproduced or used in any manner whatsoever without the express written permission of the publisher except for the use of brief quotations in a book review.

Welcome to the exciting world of tornadoes! Have you ever wondered what a tornado is and how it forms? In this book, we will learn all about tornadoes and the different types of tornadoes.

WHAT IS A TORNADO?

A tornado is a type of storm that forms when warm and cold air meet. The warm air rises, and the cold air sinks, creating a spinning motion in the atmosphere. This spinning motion can cause a funnel-shaped cloud to form, which is called a tornado.

HOW DOES A TORNADO FORM?

Tornadoes typically form when there is a lot of heat and humidity in the atmosphere. This creates an unstable atmosphere, which can lead to thunderstorms. When a thunderstorm forms, the warm air rises, and the cold air sinks, creating a rotating column of air. If the winds in this column of air become strong enough, a tornado can form.

SUPERCELL TORNADOES

These are the most common type of tornado and can be some of the largest and most destructive. They form within supercell thunderstorms, which are large, rotating thunderstorms that can last for several hours. Supercell tornadoes are often visible as a large, rotating column of clouds that extend down from the thunderstorm.

LAND-SPOUT TORNADOES

Unlike most tornadoes that form from the clouds down, land-spout tornadoes form from the ground up. They are typically weaker than other types of tornadoes, but they can still cause damage.

MULTIPLE VORTEX TORNADOES

These tornadoes are characterized by several small, swirling vortices within the larger circulation. The vortices can move independently of each other, and they can cause the tornado to change direction quickly.

GUSTNADOES

These tornadoes are typically small and short-lived, forming at the leading edge of a thunderstorm where there is a sharp boundary between the cool downdrafts of the storm and the warm, humid air outside of it.

WATERSPOUT TORNADOES

These tornadoes form over water and are typically weaker than land-based tornadoes. They can move onto land and cause damage, but they often dissipate before reaching land.

FIRE TORNADOES

These tornado-like vortices form during wildfires, often caused by the intense heat and strong winds of the fire. They can be very dangerous and can cause the fire to spread quickly.

DUST DEVIL TORNADOES

These are small, weak tornadoes that form on hot, sunny days when the ground is very hot and can create an updraft of warm air. They are often visible as a swirling column of dust or debris.

STAYING SAFE DURING A TORNADO

Tornadoes can be very dangerous, so it's important to know how to stay safe. If you hear a tornado warning, you should seek shelter in a sturdy building or underground. Avoid windows and cover yourself with blankets or pillows to protect yourself from flying debris.

Now you know all about tornadoes and the different types of tornadoes. Remember to always stay safe during a tornado, and never underestimate the power of these amazing natural phenomena!

ABOUT THE AUTHOR

My name is Nicholas Joseph David Sims. I am five years old. I have two dogs one named Mercury and the other named Bolt. Sometimes I call Bolt Venus as his nick name. My nick name is Nicholatte. My birthday is December 2nd, 2017. I started reading when I was two years old. I like to learn about the weather like tornadoes, earthquakes floods, and hurricanes. I want to be an Astrogeologists when I grow up. I really love science. I would like to thank my Dad and Mom for being the best parents in the world.

www.ingramcontent.com/pod-product-compliance
Lightning Source LLC
Chambersburg PA
CBHW042251100526
44587CB00002B/100